YOUR KNOWLEDGE HAS VALUE

- We will publish your bachelor's and master's thesis, essays and papers

- Your own eBook and book - sold worldwide in all relevant shops

- Earn money with each sale

Upload your text at www.GRIN.com and publish for free

Bibliographic information published by the German National Library:

The German National Library lists this publication in the National Bibliography; detailed bibliographic data are available on the Internet at http://dnb.dnb.de .

Imprint:

Copyright © 2017 GRIN Verlag, Open Publishing GmbH
Print and binding: Books on Demand GmbH, Norderstedt Germany
ISBN: 9783668568396

This book at GRIN:

http://www.grin.com/en/e-book/378240/laser-advanced-physical-practicum

Moritz Lehmann, Niklas Stenger

Laser. Advanced Physical Practicum

GRIN Publishing

Physikalisches Praktikum für Fortgeschrittene

Universität Bayreuth

Laser

Group 9

Moritz Lehmann, Niklas Stenger

30.08.2017

Table of contents

1. Introduction/Motivation

Lasers are very commonly used in everyday life, e.g. in laser pointers, CD-Players or the lasers used in cash registers. But what makes lasers special and distinguishes them from other light sources? Which advantages do they have and what are their applications in physics?

In the experiment "Laser" in the Advanced Physical Practicum the task is to find out which mathematic coherences there are between parameters of the laser and to test how good they match the experimental observation.

In general, what makes a laser special?

Most important are two points: an unequaled high intensity as well as a very sharp frequency range of the emitted light. Additional laser light is very sharply focused, which leads to the already mentioned high intensity of laser radiation throughout a relatively low power output, as we talk in dimensions mostly smaller than 100 mW, which is really small compared to a light bulb with 60 Watt. Nevertheless the intensity of laser light is much higher, which makes them outstanding.

Short overview of theory

2. Laser medium

2.1 Requirements for lasing

A laser is an abbreviation for "light amplification by stimulated emission of radiation". The main operating principle of a laser is to amplify a parallel beam of monochromatic light. To fulfil this task, three essential components are needed.

At first, a gain medium is needed, in which photons can be emitted by niveau changes of atoms and molecules. This only works, if more particles are in an energetic higher state than in a lower, otherwise no photons will be emitted by the laser.

Because in the thermodynamic balanced case it's quite rare that higher energy levels are occupied according to Boltzmann, in a laser an optical pump is needed to guarantee an inverse population. This is possible by pumping energy in the gain medium. Further, this only can happen, if there are more than two energy levels, because by assuming only two energy levels and an inverse population, the Einstein coefficients will mismatch in a way that it will be more likely for an electron to lose energy by stimulated emission than to gain energy by photon absorption. Additionally, spontaneous emission will make it even worse.

Also a resonator will be needed to determine which of the generated photons can leave the laser in a way that most of the photons with odd frequencies leave as the laser beam, but most of the photons with the right amount of energy stay in the resonator to establish an even more stimulated emission. This is the reason why lasers have an outstanding sharp spectral range.

2.2 Possibilities for energy levels

Because a laser with two energy levels is not possible, the next idea will be one with three energy levels. Such a laser will be possible, if the pumping process goes from the lowest to the highest energy level, and the transfer from the highest level to the middle has to happen much faster than the one from the middle energy level to the lowest in order to make the stimulated emission caused by the pumping process disappear from the rate equations. It also appears within more detailed calculations that such a three-level laser can work if the right materials and intensities are chosen. Also a high light intensity inside the resonator is required, which means that the light intensity inside the resonator has to be even higher than outside of the resonator. In this case, inverse population is possible.

For a laser with four energy levels, the needed inverse population is satisfied automatically, which can be seen by calculating the rate equations. Here the pumping process has to go from the lowest energy level to the highest one.

3. Laser resonator

This part of a laser will be realized by two mirrors with matching parameters. In our case we will use a confocal construction, which means putting two spherical mirrors in a distance at which their focal points will match in one point. The resonator therefore is used to lengthen the way through the gain medium and to sort out frequencies, since only the frequencies which match the condition for constructive interference $L = \frac{\lambda}{2} \times N$ will be amplified, where N has to be a positive integer. All other frequencies will disappear due to destructive interference. Since we talk about the interference between the incoming and reflected wave, the result will be a standing wave. The quality of the resonator has to be especially good in case the amplification of the gain medium is low. The laser resonator has to match the stability condition

$$0 < g_1 g_2 < 1; \ g_i = 1 - \frac{L}{R_i}, i = 1,2 \ (3.1)$$

L represents the length of the cavity and R is the radius of the mirrors.

We also have to differ between two kinds of modes.

The longitudinal mode is an oscillation lengthwise to the direction of light propagation, and due to the conditions at the edge of the resonator it is necessary that only special frequencies can be in the longitudinal mode.

The envelope of the amplification usually is a Gaussian curve, but since only the frequencies which matching the boundary conditions will be amplified, one will see peaks in the amplification. These peaks are the laser modes.

The envelope is generated by the Doppler broadening of the lines, which is a result of Browns movement of small particles. The spectral distance of the modes can be calculated using the formula $\Delta \nu = \frac{c}{2L}$. The longer the cavity is, the lower the spectral distance between two longitudinal modes will be.

If a wave doesn't fill the room perpendicular to the direction of light propagation, there will be a difference in the length of cavity and therefore the frequency will change. This leads to transversal modes. They can occur as TEM-Modes if using planar mirrors, otherwise hybrid modes will occur.

4. He-Ne laser

This laser uses a mix of Helium and Neon gas in a thin glass capillary. Helium is used for pumping, while Neon is the gain medium. The relation between Helium and Neon determines the wavelength of the laser. The voltage, which is necessary for the ignition of the laser, is about 15 kV. To inverse the population, electrons bump inelastically with the helium atoms in order to excite them. The following factors can influence the observed bandwidth:

First, Doppler broadening causes an increase in bandwidth, as the center of mass of a mixture of to different gases is always in motion. Additionally, the light will be pressure broadened which is a result of the low gas pressure. This happens when inelastic bumps interrupt the stimulated emission.

An advantage of this kind of laser is a quite low bandwidth of gain, allowing more modes to oscillate at the same time within the resonator. Also with the He-Ne laser a variety of different wavelengths can be emitted, which depends on the partial pressures of both gases.

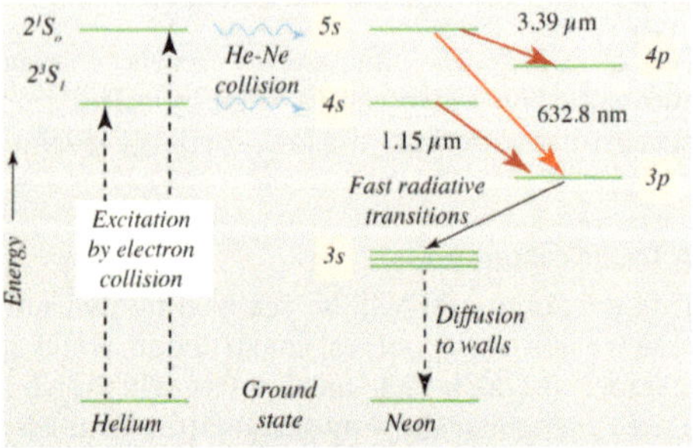

1 Energy levels for Helium-Neon-laser

In the diagram the principle for the energy levels is shown. As can be seen in above, Helium will be excited in the 2s-state and then can bump with Neon, which will drop the energy levels from 5s to 4p, or from 4s to 3p. The last one is the visible transfer, which will be used in the practicum.

5. Fabry-Perot interferometer

This tool consists out of two mirrors, which reflect the main percentage of light intensity. A small percentage is being transmitted through one of the mirrors. In this way, some proportions of the incoming beam will pass the etalon without being reflected, others will leave after being reflected several times back and forth.

In the end, all beams coming out of the mirror will be collected by a lens and projected on a screen.

The finesse is an important quantity describing the quality of a Fabry-Perot etalon, as it describes how easy it will be to separate two nearby wavelengths.

The higher the finesse of a Fabry-Perot etalon is, the sharper the interference rings on the screen will be.

This construction has similar to a laser resonator certain wavelengths with maximal intensity, caused trough interference. The single maxima also aren't very sharp like a d-distribution, but have a Gaussian curve with its full width at half maximum (FWHM). The finesse now equals the relation of the spectral range between two maxima in the intensity to the FWHM of one maximum.

$$F = \frac{\Delta\lambda}{d\lambda}$$

Here is Dl the spectral range and dl the FWHM.

The Finesse is determined by the reflectivity of the two mirrors.

$$F = \frac{\pi\sqrt{R}}{1-R} \quad (5.1)$$

6. Dielectric mirrors

In order to achieve extremely high reflectivity of the mirrors, metallic mirrors will be unsuited. Instead, we use dielectric mirrors, which consist of many layers with different refractive indices. In this mirror, also known as the Bragg mirror, again interference is used in the way that the difference in the optical path length differs by one wavelength in the different layers. This can lead to very high reflectivity and therefore a high finesse of the Fabry-Perot etalon.

7. Basic optical formulas

7.1 Snell's law

Snell's law describes the angle of light reflection when hitting an interface of two materials with different refractive indices. It follows from the principle of Fermat, which states that small changes in the way of light should cause no difference in the optical path length, which is equivalent to the formulation that light takes the shortest possible way. If deriving the time light takes between two points over an interface and setting this expression equal to zero, Snell's law can be derived. Written in a formula, it says $n_1 \sin\alpha = n_2 \sin\beta$ where a and b are the angles perpendicular to the plane of the interface.

7.2 Fresnel equations

The Fresnel equations can be derived by the boundary conditions and conservation of energy for the components of the electric and magnetic field:

a) for the parallel components:

$$\frac{E_r}{E_e} = \frac{n^2\cos\varphi - \sqrt{n^2 - \sin^2\varphi}}{n^2\cos\varphi + \sqrt{n^2 - \sin^2\varphi}}$$

$$\frac{E_t}{E_e} = \frac{2n\cos\varphi}{n^2\cos\varphi + \sqrt{n^2 - \sin^2\varphi}}$$

b) for the perpendicular components:

$$\frac{E_r}{E_e} = \frac{\cos\varphi - \sqrt{n^2 - \sin^2\varphi}}{\cos\varphi + \sqrt{n^2 - \sin^2\varphi}}$$

$$\frac{E_t}{E_e} = \frac{2\cos\varphi\sqrt{n^2 - \sin^2\varphi} - 2\cos^2\varphi}{n^2 - 1}$$

f is the angle between the incoming/reflected light ray and the surface and n is the relative refraction index which equals n_2/n_1. [1]

7.3 Brewster angle

In the Brewster angle, no p-polarization passes the interface. The reflected light has linear polarization, because only the s-polarized parts are being reflected. If combining Snell's law with the condition that no light is reflected, one gets

$$\theta = \arctan\left(\frac{n_2}{n_1}\right). \quad (7.3.1)$$

To determine the Brewster angle a polarization filter can be used, so that the reflected light is only s-polarized if light hits the interface under the Brewster angle. Therefore, if setting the polarization filter at zero degrees, no light will be transmitted. If now measuring the angle under which light hits the plate, the Brewster angle is being observed. For a glass plate this angle should be around 55-60 degrees.

8. Propagation using ray transfer matrix analysis

If calculating light transfer through different components, the calculations can be simplified if using paraxial approximations. One can use matrices to describe the light transfer by using the position and angle respectively slope of the incoming beam as components of the initial vector. By multiplying this expression with a ray transfer matrix, the outgoing light ray can be determined. Generally, this matrix form will look like

$$\begin{pmatrix} x_o \\ x_o' \end{pmatrix} = \begin{pmatrix} A & B \\ C & D \end{pmatrix} \begin{pmatrix} x_i \\ x_i' \end{pmatrix}$$

Here, the index "i" stands for the incoming ray and "o" for the outgoing one. The expressions with a stroke stand for the slope of the rays.
Below some examples for such matrices are listed.

a) for translation

$$\begin{pmatrix} 1 & d \\ 0 & 1 \end{pmatrix}$$

b) for a thin lens

$$\begin{pmatrix} 1 & 0 \\ -\dfrac{1}{f} & 1 \end{pmatrix}$$

c) for transmission through a dielectric with refractive index[2] n

$$\begin{pmatrix} 1 & \dfrac{d}{n} \\ 0 & 1 \end{pmatrix}$$

To get the matrix for different systems, the matrices have simply to be multiplied.

9. Properties of M^2

M^2, also known as the beam quality factor represents how easy the laser beam can be focused. With increasing beam quality factor, it will be harder to focus the light. The optimum is a beam quality factor equal 1, which is satisfied for a Gaussian beam. In reality, this quantity will always be greater than one, leading to lower beam quality as it decreases by one over the square of the beam quality factor. For the determination of M^2, the middle section of a laser ray can be measured by calculating the variance of the intensity using the known expressions[3]

$$\langle x^2(z) \rangle = \frac{\int (x - \langle x \rangle)^2 I(x,y,z)\,dxdy}{\int I(x,y,z)\,dxdy} \qquad \langle y^2(z) \rangle = \frac{\int (y - \langle y \rangle)^2 I(x,y,z)\,dxdy}{\int I(x,y,z)\,dxdy}$$

The lower the variance and the standard derivation is, the better (more like a Gaussian beam) will be the laser.

10. Coherence

Interference between waves requires them to be coherent, which means that they share the exact same frequency apart from a phase lag. If the waves aren't a perfect sine signal, the difference in the path length can not be arbitrary, but a maximum will exist which must not be exceeded, since otherwise no interference can be detected. This maximal difference in the path length between two waves is called coherence length. This phenomenon is based on spatial coherence, which is the decisively point when two waves are separated in space.

If the wave is not perfectly periodic, but there are greater periods in which the wave swings quite predictable, the time of a period will be called coherence time.

Protocol of the experiment

11. Parameters of the practicum

Date: 31.08.2017
Time: 8:15-16:00
Room: B 11, 0.05, "Fortgeschrittenenpraktikum Laser"
Measurements: Moritz, partly Niklas
Protocol: Niklas
Pictures of the experiment: Moritz
Used tools:　　Polytec laser (as help laser)
　　　　　　　Melles Griot laser
　　　　　　　THORLABS (dielectric mirrors)
　　　　　　　Pictures taken with the camera of an iPod touch
　　　　　　　Rigol DS 1054 oscilloscope

12. Autocollimation of the laser beam

To make the laser ignite, we need to adjust our beam properly. Adjusting the dielectric mirrors is done by using a second laser. First, we try to get a sharp point at the paper, which represents the screen. Then first the dielectric mirrors are inserted, the one with a reflectivity of 99,9 %, and then the one with 98% reflectivity. To ensure that the beam is fixed properly, an aperture is inserted in the beam to block a part of the radiation of the help laser. After a few failed attempts, the laser ignites and the adjustment laser can be switched off. Also, the aperture can be taken out of our experiment.

The distance between the two dielectric mirrors must be between 50cm and 100cm. For the first experiment, we choose about 52cm.

The ignition of the laser can be seen as red spot at a screen at the other side of the room.

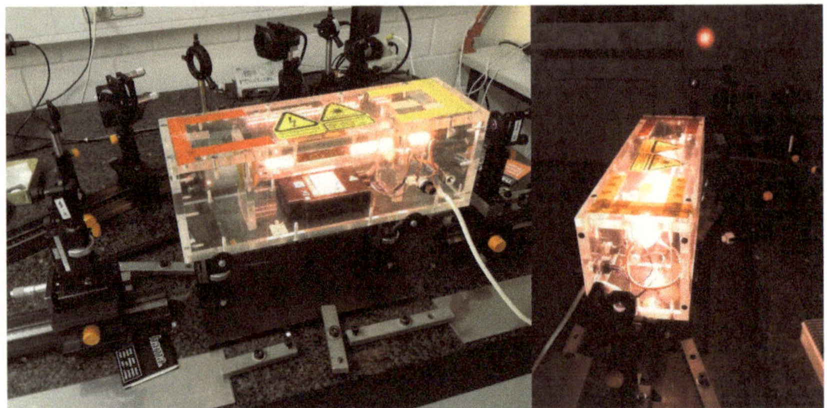

3 Construction of the experiment, from the side	**2 Projecting the laser beam at a screen**

13. Transversal modes

By putting a thin wire inside the cavity and by making little adjustments to the mirrors, different transversal modes can be seen. Astonishing to see is the extremely high sensitivity of the adjusted mirrors. Even if someone walks on the ground, the pattern on the wall moves. We also put a lens into the beam to enlarge the patterns of the transversal modes. It's important to notice that the $TEM_{a,b}$ mode means a+1 maxima in the x-direction and b+1 maxima in the y-direction, since the $TEM_{0,0}$ mode is circular shaped, which has already one maxima in both directions.

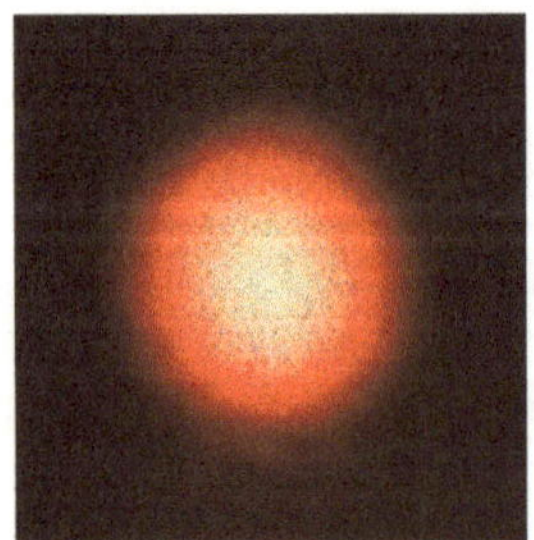

4 $TEM_{0,0}$ mode

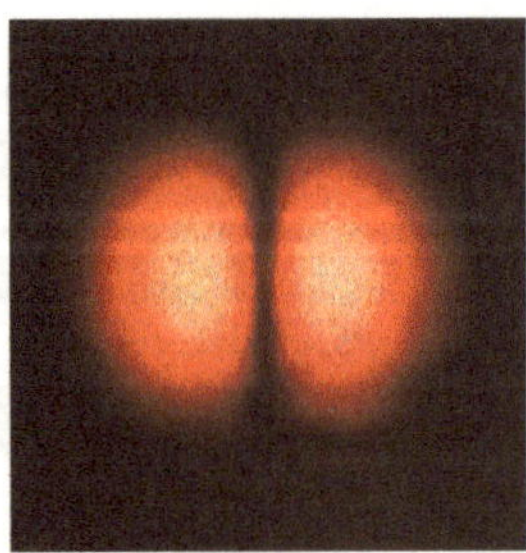

5 $TEM_{1,0}$ mode

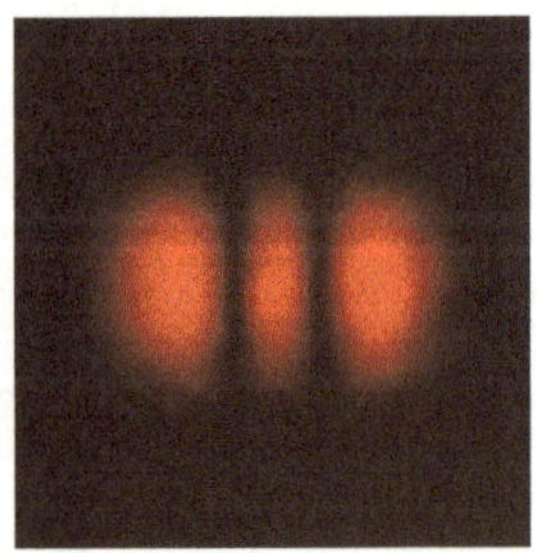

6 $TEM_{2,0}$ mode

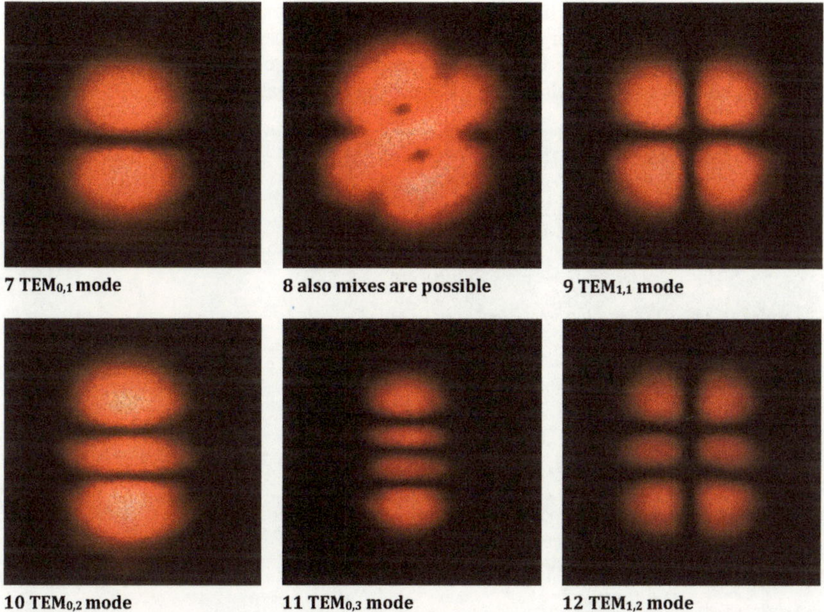

7 TEM$_{0,1}$ mode	**8 also mixes are possible**	**9 TEM$_{1,1}$ mode**
10 TEM$_{0,2}$ mode	**11 TEM$_{0,3}$ mode**	**12 TEM$_{1,2}$ mode**

14. Gain factor and Brewster angle

In this part the task is to measure the Brewster angle and to calculate the gain factor of the laser medium. The evaluation is supported by a computer program, which calculates the intensity of the light by measuring the voltage generated in a photo diode. The light is being reflected by a mirror into the diode and by optimizing the distances and angles of the mirrors maximal intensity shall be achieved. To control this point, we watch the incoming and reflected beam at the mount of the mirror and match them. For the angle, at which both beams match, we get an offset at the computer system from -13,3 degrees, which has to be taken into our calculations. Because Brewster angle is expected by about 56 degrees due to the roughly known refractive index of the glass, we make a measurement from 23 to 63 degrees. The Brewster angle is expected to be at 43 degrees in the system of the computer program. First, we make a few quick test measurements with a step from 0,5 degrees and try to optimize the lasing, since for angles larger than 50 degrees hardly oscillation is seen. Then we started an exact measurement with a step from 0,02 degrees. It is important to say that the computer program takes 20 measurements for each point and calculates the mean value and standard derivation; this will be useful for the calculation of the error later.

15. Axial modes

In this measurement, an oscilloscope is used to calculate the distance between two modes and the width of a single mode. Later an interferometer will be inserted in the construction to measure the free spectral range, the full width half maximum at one peak and therefore the finesse of the interferometer.

Calibrating our oscilloscope, we get a time difference from about 1,7 ms; which would lead to 170 MHz frequency, which is too far away from the expected 250 MHz, so we had to repeat this experiment later and carry on with the other tasks.

After an improvement made by Uwe Gerken, we get a good-matching value for our frequency, which can be calculated over the width of one peak.

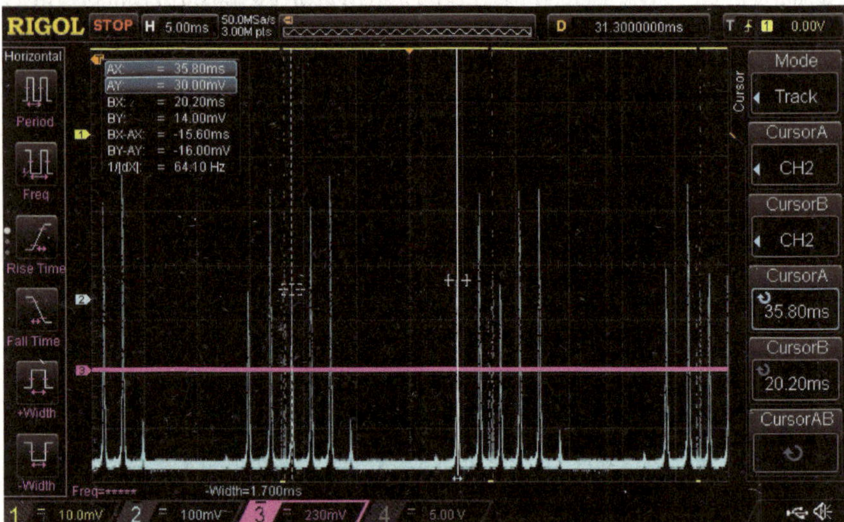

13 Measurement for the axial mode, 15,6 ms equals 2 GHz.

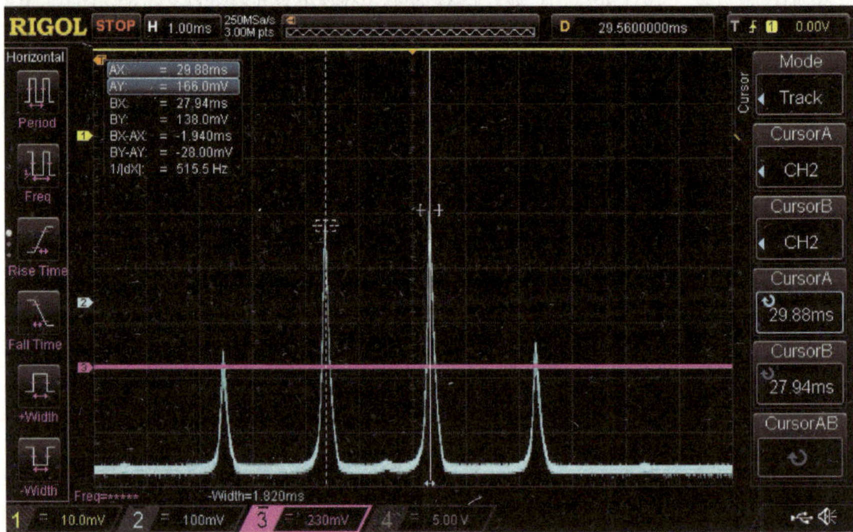

14 Difference between two peaks, 1,94 ms equals 249 MHz.

Now, the interferometer is used to get a single-mode-state of the laser.

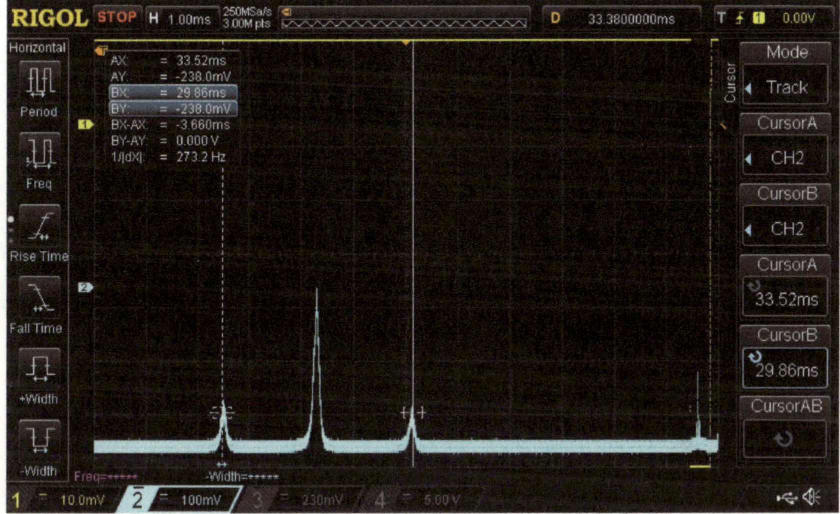

15 Single mode state of the laser with few perturbations

The frequency of the peak can be averaged out of the two markers. We get 31,69 ms, which equals (with 15,6 ms = 2 GHz) 4,06 GHz. The difference between two peaks equals 1,83 ms, which is equivalent to 234 MHz. For the last value it should be similar to the 249 MHz from the measurement done before, because the difference between two peaks isn't influenced by their height. The peak frequency should equal to an integer of the axial mode because of the boundary conditions, which is here quite good satisfied (4 GHz should be expected).

16. Holography

To do this experiment, the laser ray has to be reflected and diverged by a lens to enlarge the holography and then the light hits a holographic plate, where the object can be seen in three dimensions. The coherence length is calculated to

$$L = \frac{c}{\delta v} = \frac{3,00 \cdot 10^8 \, \frac{m}{s}}{17,9 \, MHz} = 16,8 m$$

With an error of 1/20 of one div, which here is 1/20 of 200 μs, therefore 10 μs equalling 1,3 MHz for the frequency, we get

$$s_L = \frac{c}{(\delta v)^2} \cdot s_{\delta v} = 1,2 m$$

using the error propagation. For the coherence length, we get a result of L=16,8±1,2m. We expected a value from larger than roughly 5 meters (distance between laser and holographic plate), because if coherence is not fulfilled in the holographic plate, no holographic pattern will be seen.

16 Holographic pattern

17. Gaussian laser profile

For this experiment, we use the camera CCD and to get the right intensity on the CCD we looked at the computer output and used filters. After a bit of trying we find out which filters to use, namely which AU numbers, where one AU means a transmission by 10^{-1}, which equals 10% of the incoming light. The CCD will now be put in the focus of the laser ray and moved back and forth to find out the waist of the beam. This can be used to calculate the beam quality factor M^2. Because only the full width at half maximum is necessary to calculate, the absolute value of the maximum doesn't matter. Therefore the used filters aren't needed for our calculations. Below two pictures are shown as an example.

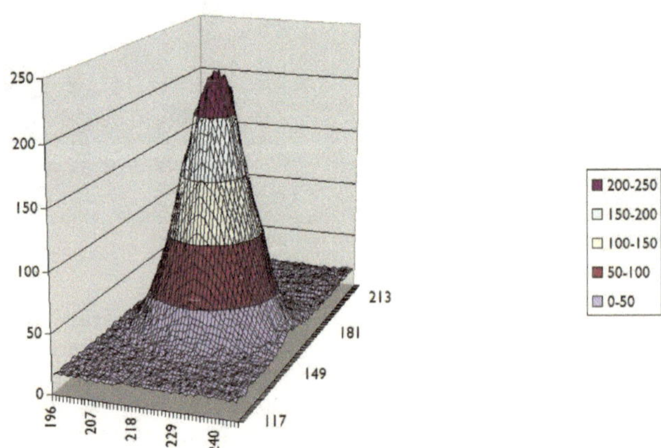

17 Situation when the CCD is in the focus of the lens

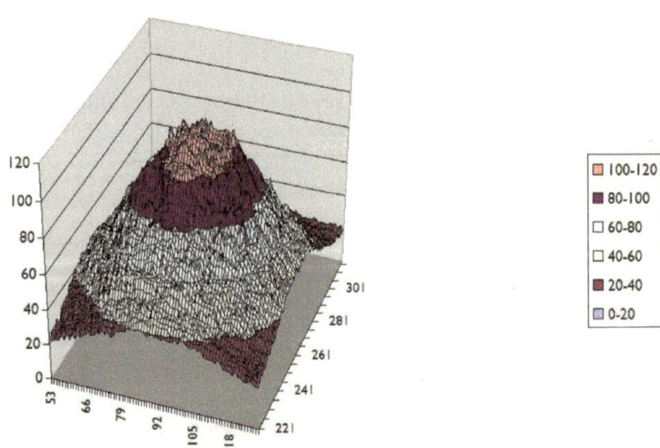

18 Situation when the CCD is after the focus of the lens

No errors are marked in the sketch, because it will be confusing to have them for more than 2000 points.

Evaluation of the experiment

To determine the errors of the measurement, we observe that the error in reading off the displays is much larger than the systematic errors, for which we also didn't find any values. The values for the errors which we estimated are also given in the protocol.

18. Identification of the transversal modes

The identification has already been done in the upper part, so here let us note also hybrid modes are possible. The state shown in picture 7 is a hybrid mode out of the $TEM_{1,1}$ and the $TEM_{1,1}$ rotated by 90 degrees. The mode could be called $TEM_{1,1}^*$. To answer the question, why the mirrors are put up in the confocal collocation to make it easier to excite higher transversal modes, we can say that higher modes mean higher energy levels. Because the laser has to get maximum energy to get higher modes it must be hit be both rays from both sides of the resonator with maximal intensity, which is ensured, if the middle if the lasers is in the focus of both mirrors. In this case, the probability for the excitement of higher modes is maximal. If moving a mirror out of this position, the energy will drop and so will the probability for the excitement of higher modes.

19. Calculation of gain factor and Brewster angle

At first, let's have a glance at the data plot before going into the data itself.

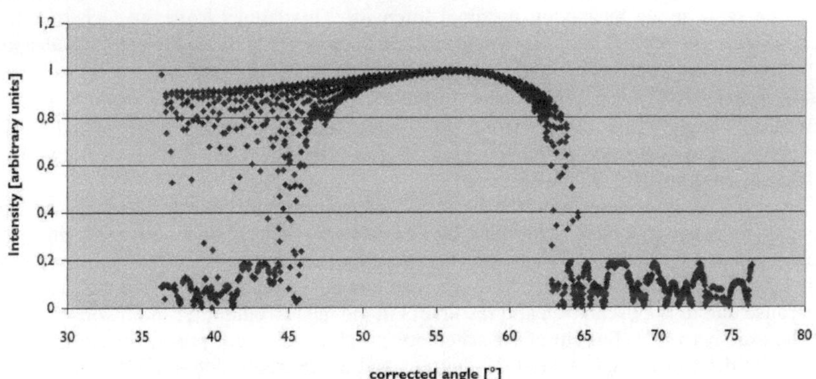

19 Intensity of the laser during rotation of the glass plate

In the figure above the offset of 13,3 degrees is already corrected.
In this sketch, the Brewster angle seems to be at roughly 53 to 58 degrees. We can improve this sketch by using a smaller angle range.

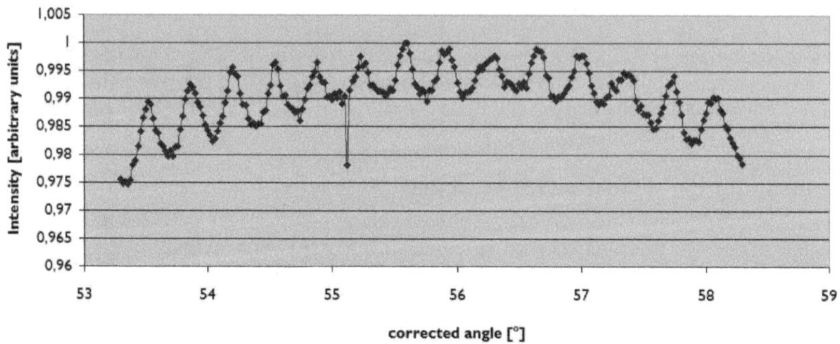

20 Fine figure from fig. 18

In this sketch can be seen, that the Brewster angle will lie roughly at 55,5 degrees. But it will be even better, if the angle will be read out of the data. This will be done below.

The intensity oscillates because the plate isn't infinitely thin, but has a certain width (which will be calculated later). This leads to reflection at both interfaces, and both rays will interfere. Because the wavelength is much smaller than the expected width of the glass plate, the oscillation will be quite fast.

It can be seen that the intensity is asymmetric in relation to the Brewster angle, which can be seen above by having maximal intensity. This comes from the fact, that the Brewster angle doesn't equal zero degrees, which means that in the Fresnel formulas for the transmission and reflection for symmetry would have the condition $I(\alpha_B + d\alpha) = I(\alpha_B - d\alpha)$. But these formulas aren't symmetric in relation to the Brewster angle. This comes from the cosine and sine expressions, which have symmetries to angles like 90 or 0 degrees. Because the Brewster angle is at roughly 56 degrees, no symmetry will be observed.

The next task is to determine the Brewster angle (maximal intensity) and the critical angle (no intensity). This is be done by analyzing the data shown above. Maxima are observed at 42,28 and 42,3 degrees. The value doesn't matter, because it is normalized at 1 for the maximal intensity. Now we read out the critical angles. This isn't as easy, because due to the oscillation and the errors in the meausrement it can't be observed to drop exactly to zero. But out of the schematic graph given in our paper, we can estimate that the drop at the left side at 32 degrees and at the right side at 50 degrees is the critical angle. Also, as can be seen that the measurement at angles larger than 50 degrees, the signal is not very steady, maybe because of perturbation caused by the light in the environment of the experiment. Exact values for the critical angles are 32,18 degrees and 50,32 degrees, which are read off the data in the .dat data file. To estimate the error, we take into account that the read error equals one smallest step length, which here means 0,02 degrees (step of the measurement). It seems that measurement of the intensity is quite small, because one sharp dip is recognized, so we estimate here 0,02 degrees, too.

Furthermore, in the measurement of the offset a difference from 0,03 degrees wasn't distinguishable, so we have to calculate with that if calculating absolute values of the angles. This means for the Brewster angle an error of $s_B = \sqrt{0{,}02^2 + 0{,}02^2 + 0{,}03^2} = 0{,}04°$.

This leads to a Brewster angle of <u>55,59±0,04 degrees</u>. For the glass plate, values around 55-58 degrees are typical, so our result seems to match the expectations.

For the critical angles, the error stays the same, so we have angles of <u>45,48±0,04 degrees</u> and <u>63,62±0,04 degrees</u>. The critical angle is observed when total reflection happens. An expected value can be calculated if inserting an assumption for the reflective indices of the glass plate. If using $\theta = \arcsin(n)$, which is derived from Snellius law and analyzing, when it doesn't fulfil the transmission condition and n=1,5 (a typical value for glass), we get 41,8 degrees, for a refractive index of 1,4 45,6 degrees. Therefore our result matches the expectations. We can also calculate the critical angles in relation to the Brewster angle, where we can take offset out of the error, so we'll get an error of $s_{rel} = \sqrt{2} * 0,02 = 0,03°$. For the relative values, we get <u>-10,11±0,03 degrees</u> and <u>8,03±0,03 degrees</u>. This also shows the asymmetric behaviour of the function, since both values aren't the same.

Now the win-loss-balance for the intensity is calculated. We call the gain factor in the laser medium v and the transmission coefficient trough the glass plate T. T will depend on the angles and the width of the plate, which will be calculated later. We get the following equations for the intensities:
$$I_1 = vI_0 \Rightarrow I_2 = vTI_0 \Rightarrow I_3 = VTR_2I_0 \Rightarrow I_4 = VT^2R_2I_0 \Rightarrow I_5 = V^2T^2R_2I_0 \Rightarrow I_6 = V^2T^2R_1R_2I_0.$$
To get these equations, we can follow the beam trough the resonator. Every time the beam is reflected (or transmitted), the intensity will drop at the factor R (or T). If the beam goes through the gain medium, the Intensity will raise by V (gain factor). Both mirrors have different reflection coefficients $R_{1/2}$.

I_6 has to equal I_0. The task will be to calculate T.

For an interface we will use the relation

$$\frac{A_r}{A_0} = \frac{\tan(\alpha - \beta)}{\tan(\alpha + \beta)} \quad (19.1)$$

To calculate the relation between a and b, the refractive index of the glass will be needed, which can be calculated by the Brewster angle. Starting point is the equation $\theta = \arctan(\frac{n_1}{n_2})$. If assuming that the refractive index of air equals 1, which can be said when regarding the errors of the calculations. In this way, we get $n_1 = \tan\theta = 1,460$ and $s_n = 0,004$. We used here n_2 as the refractive index of air, which can be set to one and n_1 equals the refractive index of the glass plate.

With the law of Snellius $n_1 \sin\alpha = n_2 \sin\beta$. We derive

$$\sin\beta = \frac{1}{n_2}\sin\alpha \Rightarrow \beta = \arcsin(\frac{1}{n_2}\sin\alpha) \quad (19.2)$$

This can be inserted in the relation given above, so we get

$$\frac{A_r}{A_0} = \frac{\tan(\alpha - \arcsin(\frac{1}{n}\sin\alpha))}{\tan(\alpha + \arcsin(\frac{1}{n}\sin\alpha))} \quad (19.3)$$

Here, we renamed the variable n_2 with n. This is the equation for the first interface. For the second interface, we have to calculate with new angles for the amplitude.

Both waves will interfere. Because the angle under which the beam leaves the first interface equals the angle under which light hits the second interface and light will leave the plate under the same angle like the incoming angle due to equal refractive indices, we simply can switch b and a in the equation above, which simply switches the sign of the amplitude. But this wave has a different phase. Since the waves start from other edges of the plate, they will interfere. Let's call the angle of the plate g.

The length trough the plate equals $l = \dfrac{d}{\cos\gamma}$. Further, $\gamma + \alpha = 90°$ has to be satisfied following from geometric considerations. Therefore we calculate the phase difference between the two reflected waves. It is equal to $2\pi\dfrac{l}{\lambda}$. Only the integer part of the phase difference determinate the amplitude of the wave. The superposition of both waves will be maximal for a phase difference of 180 degrees (!), because both waves have different signs. It means we can use

$$A = A_{max}\cos(2\pi\frac{l}{\lambda} + \pi)$$

The maximal amplitude satisfies the equation

$$\frac{A_r}{A_0} = \frac{2\tan(\alpha - \arcsin(\frac{1}{n}\sin\alpha))}{\tan(\alpha + \arcsin(\frac{1}{n}\sin\alpha))} \quad (19.4)$$

Because this is the reflected amplitude and the energy conservation has to satisfy the equation

$$\frac{A_t}{A_0} = 1 - \frac{2\tan(\alpha - \arcsin(\frac{1}{n}\sin\alpha))}{\tan(\alpha + \arcsin(\frac{1}{n}\sin\alpha))} \quad (19.5)$$

If inserting the equations, we can derive the equation:

$$A = (1 - \frac{2\tan(\alpha - \arcsin(\frac{1}{n}\sin\alpha))}{\tan(\alpha + \arcsin(\frac{1}{n}\sin\alpha))})\cos(2\pi\frac{d}{\cos(90° - \alpha)} + \pi)A_0 \quad (19.6)$$

Therefore for the transmission coefficient follows:

$$T = 1 - \frac{2\tan(\alpha - \arcsin(\frac{1}{n}\sin\alpha))}{\tan(\alpha + \arcsin(\frac{1}{n}\sin\alpha))})\cos(2\pi\frac{d}{\cos(90° - \alpha)} + \pi) \quad (19.7)$$

This equation was derived out of the interference of the reflected beams at both interfaces and using the equations derived above and the energy conservation.
This equation depends on d and a. We can insert this equation in the win-loss-equation. $I_6 = V^2T^2R_1R_2I_0$ brings us to

$$V = \frac{1}{T}\sqrt{\frac{I_6}{R_1R_2I_0}} \quad (19.8)$$

We insert an angle and use $I_0=I_6$. We have to use the point with the maximal intensity, which is the Brewster angle. The width of the plate can be calculated by counting the minima in a defined angle. We count 28 in 480 data points. This means 28 in 9,6 degrees. We calculate the total minima in 360 degrees. It follows

$$N = \frac{360}{9,6}*28 = 1050$$

maxima. That means, the width of a plate is 1313/2p times the wavelength. The wavelength equals 632,8 nm. Therefore the width of a plate equals

$$d = \frac{1050}{2\pi}*632,8nm = 106\,\text{mm}$$

To get this equation, we kept in mind the interference condition. A maximum is observed when the difference in the path lengths equals an integer of the wavelength. Because the difference in the path lengths is a cosine oscillating with 2p, we have to divide the maxima through this factor. The width of the plate will go linear into the number of maxima, because the path length difference is also linear to the number of maxima. Then, each maximum equals one maximum of the cosine. The result which has been calculated seems to be in the right dimension. The error in reading out oscillations is quite large because oscillations have a long stable value. The error is estimated to be 10 data points, which is 0,2 degrees. For the error of the maxima we get

$$s_N = \frac{360*28}{9,6^2}*0,2 = 22$$

The error of the plate width is

$$s_d = \frac{22}{2\pi}\cdot 632,8nm = 2\,\text{mm}$$

It follows <u>d=106±2 mm</u>. This can be inserted in the equation for the gain factor. Now we have to insert the Brewster angle (because there is the maximum intensity) into the equation for the gain factor. In this case, the transmission factor is 93,1%. We can insert this in the equation

$$V = \frac{1}{T}\sqrt{\frac{I_6}{R_1R_2I_0}}$$

which was derived above and get v=1,086. We assume that the error of the angle measurement is tiny (computer controlled) compared with the error of the determination of the angle of the width. We have to derive the cosine part which gets us a sine. It follows

$$s_v = \sin(2\pi\frac{d}{\cos(90°-\alpha)}+\pi)*\frac{2\pi}{\cos(90°-\alpha)}*s_d = 0,001$$

Therefore is <u>v=1,086±0,001</u>. We used the error of the width of the glass plate and the law of error propagation.

20. Fabry-Perot-Interferometer

The distance of two modes and the width of a mode already have been determined in the upper part.

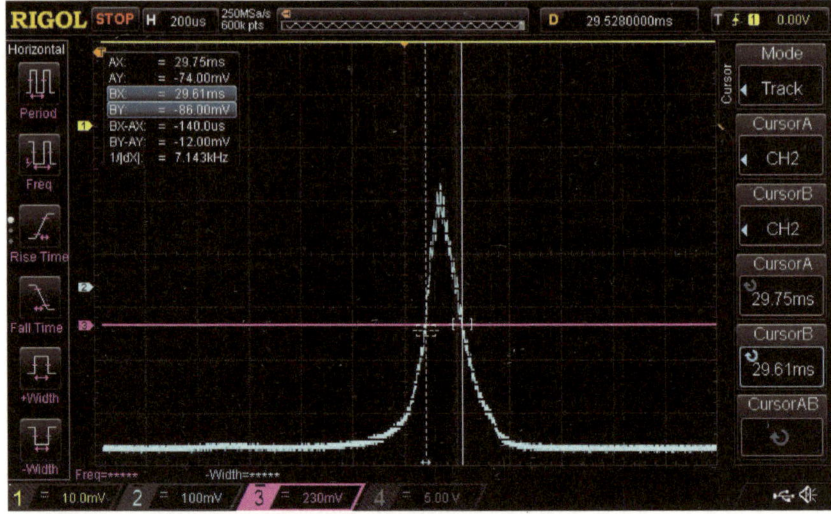

21 width of the gain profile 140 ms equals 17,9 MHz.

To calculate the finesse, we have to divide the free spectral range trough the FWHM. It follows

$$F = \frac{2\,GHz}{17,9\,MHz} = 112$$

The error in reading out time is estimated to be 1/4 of a small unit, which means 1/20 from a division. For the first picture, it means 0,25 ms, for the second picture 0,01 ms. 15,6 ms equals 2 GHz, which means errors in the frequency of 32,1 MHz and 1,3 MHz. We now need the error propagation, which provides us the following expression (we named the counter "a" and the denominator "b").

$$s_F = \sqrt{(\frac{1}{b}s_a)^2 + (\frac{-a}{b^2} * s_b)^2} = 8,3$$

Therefore the result for the Finesse is <u>111±8</u>.

The frequency of the mix can be simply read off the picture (red rectangle in the photo)

It follows with an estimation of the error <u>253,5±0,5 MHz</u> for the mixture frequency.

The task to change the cavity length fails because of too low differences which vanish in the error of reading.

To increase the precision it would be necessary to improve the measurement of the frequency.

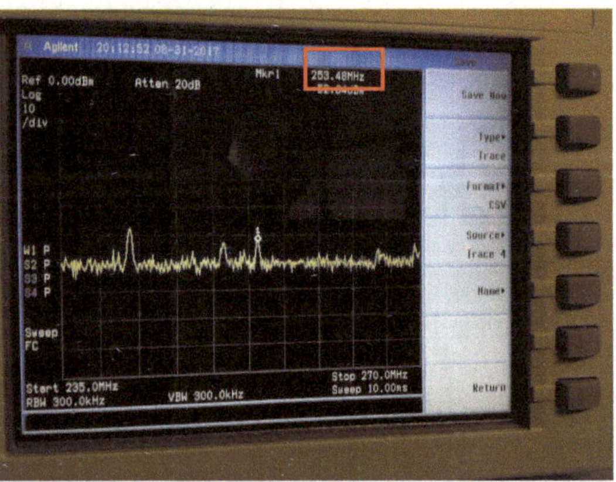

22 axial modes with fast photo diode

21. Gaussian laser profile

Now we have to determine the ray quality factor. At first, we determine the FWHMs for different distances. The determination is done by fitting a Gaussian profile at our data. This is shown exemplary for the position 5 cm before the focus. Because only one direction differs when moving along the axes, it is enough to make a fit for the x-axis. Also, it's useful only to fit the zones in which the intensity is larger than 20, because the fit should match the Laser profile, not the background noise. 8,8mm equal 640 pixels according to the received data, because in the fit our parameters will be given by pixels. With our fit, we get a standard derivation of 10,6 pixels. This equals

$$\sigma = 10,6\,px \cdot \frac{8,8mm}{640\,px} = 0,145mm\,.$$ For the FWHM we can use the relation between standard

derivation and FWHM for a Gaussian function: $FWHM = 2\sigma\sqrt{2\ln 2} = 0,35mm$.

As can be seen in the graphic below, our fit matches quite good. It can be seen that in the outer areas (larger than roughly 2 standard derivations) the fit begins to lack the estimations. This is caused by the background noise, which should not be taken within the fit.

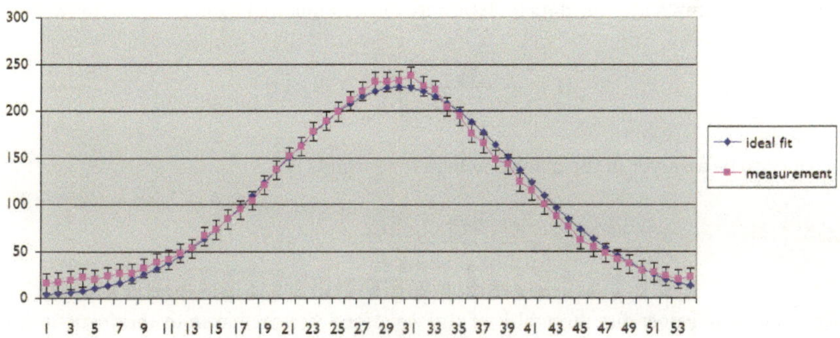

23 The fit in blue and the measurement in purple match good

focal dist	FWHM x	FWHM y	unit 0,1 mm
-20	11	10,2	-40
-15	8,4	7,8	-30
-10	6,3	5,6	-20
-5	2,9	2,7	-10
0	0,7	0,6	
5	3,5	3,2	10
10	6,7	6,2	20
15	14,2	13,6	30
20	19,3	17,6	40

24 Values calculated for the FWHMs in different distances.

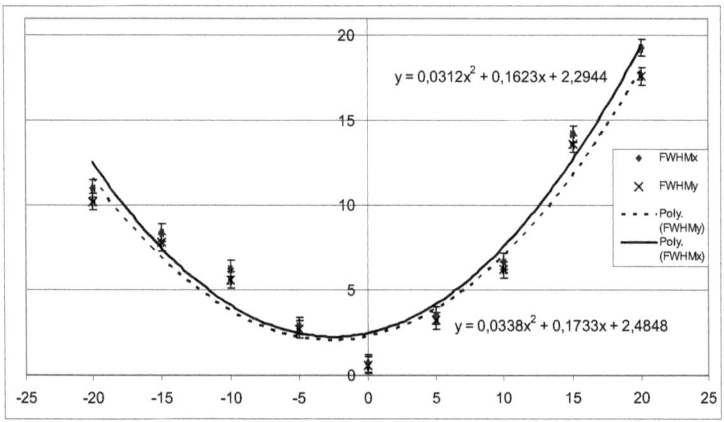

25 Diagram for dependency of FWHM from the distance from the focus.

Now, we can simply use this data to determine the ray quality factor using the formula

$$M^2 = \frac{\pi d_0^2}{2\lambda(L_0 - L_1)}\sqrt{\frac{d_1^2}{d_0^2} - 1} \quad (21.1a)$$

It seems to be an error in the manual, where the formula is

$$M^2 = \frac{\pi d_0^2}{2\lambda(L_2 - L_1)}\sqrt{\frac{d_1^2}{d_0^2} - 1} \quad (21.1b)$$

We insert our values for the width of the waist and the distances to the focus in the equation (21.1a) and therefore can calculate values for the beam quality factor. This will be done for the FWHM in x-value and in y-value. In the table below the calculated values can be seen.

focal dist/cm	FWHM x/0,1 mm	FWHM y/0,1mm	M^2x	M^2y
-20	11	10,2	0,95	0,76
-15	8,4	7,8	0,97	0,77
-10	6,3	5,6	1,09	0,83
-5	2,9	2,7	0,98	0,78
0	0,7	0,6		
5	3,5	3,2	1,19	0,94
10	6,7	6,2	1,16	0,92
15	14,2	13,6	1,64	1,35
20	19,3	17,6	1,68	1,31
d_0^2	0,7	0,6		
		average	1,21	0,96

26 Table of calculated values for the beam quality factor

It will be noticed, that the upper graph isn't symmetric to the y-axis. For the error calculation, we use the following identities:

$$s(L_0 - L_1) = \sqrt{2} \cdot s_L \tag{21.2}$$

$$\frac{dM^2}{dd_0} = \frac{2\pi d_0}{2\lambda(L_0 - L_1)}\sqrt{\frac{d_1^2}{d_0^2} - 1} + \frac{\pi d_0^2}{2\lambda(L_0 - L_1)}\frac{1}{2\sqrt{\frac{d_1^2}{d_0^2} - 1}}\frac{-d_1}{2d_0^3} \tag{21.3}$$

$$\frac{dM^2}{dd_1} = \frac{2\pi d_0}{2\lambda(L_0 - L_1)}\frac{1}{2\sqrt{\frac{d_1^2}{d_0^2} - 1}}\frac{2d_1}{d_0^2} \tag{21.4}$$

$$\frac{dM^2}{d(L_0 - L_1)} = \frac{-\pi d_0^2}{2\lambda(L_0 - L_1)}\sqrt{\frac{d_1^2}{d_0^2} - 1} \tag{21.5}$$

Now we can use the error propagation and add the errors quadratic. This is done in the table. The equation for the error is quite complex and lengthy. Because even the errors of the values of d and L have to be estimated, the error propagation will not give us a good error approximation. It's better to estimate the error out of the difference of the M^2 calculations to 5% of the value. We get for the beam quality factors: $M_x^2 = 1{,}21 \pm 0{,}06$ and $M_y^2 = 0{,}96 \pm 0{,}05$. It matches our expectations quite well, because for an ideal profile the value should equal 1, and for a laser like we used it a value slightly over 1 is expected and seems to be realistic.

22. More questions

To calculate the effective focal length we simply have to calculate the distance between lens and the focus. We get <u>350±1 mm</u> by reading out the distance from the lens, in which the waist of the focus is the smallest. We used the CCD to determine this point. This point is also the point for the measurement of d_0.
The minimal waist of the laser beam is[4]

$$w_0 = \sqrt{\frac{\lambda d}{2\pi}}$$

Inserting our data we get 0,070 mm for the theoretical value of the waist. Now let's assume that the error in the wavelength is small, because the error in d will be much greater. So we get for the error in the theoretical waist.

$$s_w = \frac{1}{2\sqrt{\frac{\lambda d}{2\pi}}}\frac{\lambda}{2\pi}s_d$$

which will be rounded to <u>0,070±0,001 mm</u>. This value matches quite well the value for the x-axis. For the y-axis, the values don't match quite well, which lies in the hybrid state out of two modes, from which one is rotated by 90 degrees. Other problems are the reading out of the full width at half maximum and perturbation light during the measurements.
To calculate the laser intensity we calculate the expression

$$I = \frac{P}{A} = \frac{P}{\pi ab} = 75{,}8\frac{kW}{m^2}$$

The error of the width is 0,3 mm, so we calculate

$$s_I = \sqrt{\left(\frac{-P}{\pi a^2 b}s_a\right)^2 + \left(\frac{-P}{\pi a b^2}s_b\right)^2} = \frac{P}{\pi a b}\sqrt{\left(\frac{s_a}{b}\right)^2 + \left(\frac{s_b}{a}\right)^2} = 49{,}9\frac{kW}{m^2}$$

The error is extremely large, since the length of the semi axes is low and the error is high. We can't calculate a result with a better precision than <u>76±50 kW/m²</u>.

23. Conclusion

This experiment provided us insight in how a laser is set up and ignited. We looked onto different modes and calculated the waist of laser beams.
Lasers can be used in many different physical applications. We furthermore used our knowledge out of the second semester in experimental physics, the optic lecture and "Messmethoden". We will use lasers on the university quite often, even in the Master, because the basic principle is quite simple - but there are many applications.

24. Bibliography

1   Dr. W. Schöpf: Practicum "Versuch Pol"
2   taken from Kogelnik, Li: "Laser Beams and Resonators", Applied Optics, October 1966
3   formulas out of Eichler, Dünkel, Eppich: "Die Strahlqualität von Lasern", www.laser-journal.de, date: 04.09.2017, S.64
4   http://www.spektrum.de/lexikon/physik/strahltaille/13982, date: 19.09.2017